CHEERS FOR CAREERS!

I WANT TO BE A CARPENTER

by Julie Murray

Cody Koala
An Imprint of Pop!
popbooksonline.com

Hello! My name is Cody Koala

This book is filled with videos, puzzles, games, and more! Scan the QR codes* while you read, or visit the website below to make this book pop.

popbooksonline.com/carpenter

*Scanning QR codes requires a web-enabled smart device with a QR code reader app and a camera.

abdobooks.com

Published by Pop!, a division of ABDO, PO Box 398166, Minneapolis, Minnesota 55439.

Printed in the United States of America, North Mankato, Minnesota.

052024
082024

Cover Photo: Shutterstock Images
Interior Photos: Shutterstock Images
Editor: Elizabeth Andrews and Grace Hansen
Series Designer: Colleen McLaren

Library of Congress Control Number: 2023947425

Publisher's Cataloging-in-Publication Data

Names: Murray, Julie, author.
Title: I want to be a carpenter / by Julie Murray
Description: Minneapolis, Minnesota : Pop!, 2025 | Series: Cheers for careers! | Includes online resources and index
Identifiers: ISBN 9781098246044 (lib. bdg.) | ISBN 9781098246600 (ebook)
Subjects: LCSH: Carpentry--Juvenile literature. | Building trades--Juvenile literature. | Construction--Juvenile literature. | Servicing trades--Juvenile literature. | Occupations--Juvenile literature.
Classification: DDC 694--dc23

Table of Contents

Chapter 1

Working with Wood

Carpenters are skilled workers. They work with wood and other materials. They do a variety of jobs in the **construction industry**.

Carpenters are in high demand in the United States.

Watch a video here!

Chapter 2

Skills Needed

Many skills are needed to be a carpenter. Attention to detail and math skills are very important. The job requires math and measuring.

Learn more here!

Carpenters work with other **tradespeople** and **clients**. They need to communicate well with others. They also need to be good problem solvers.

Carpenters use many tools, such as saws, hammers, power tools, and more. They need to make exact measurements and perform tasks with their hands. The job can be physically demanding.

clamp
chisel
hammer
tape measure
saw
combination square

Chapter 3

How to Become a Carpenter

Experience is needed to become a carpenter. This can be done through an **apprenticeship**. This is a good way to learn through hands-on experience.

An apprenticeship lasts for three to four years.

Explore links here!

Some colleges offer two-year programs in carpentry. This is another good way to gain needed skills. Learning math skills and how tools work are also important steps in becoming a carpenter.

Chapter 4

Different Jobs

Carpenters build and **install** many different things. They can work inside or outside. Some build **frames** for floors, roofs, and walls.

Carpenters also build furniture.

Complete an activity here!

They also install stairs, doors, and cabinets. They make sure that all the

materials fit together. They complete finishes such as trim and **molding**.

Being a carpenter is a rewarding job! It requires many skills to complete a project from beginning to end.

There are more than 1.3 million carpenters in the United States.

Making Connections

Text-to-Self

If you were a carpenter, what would you want to build? Why?

Text-to-Text

Can you think of another book you have read that has a carpenter building something?

Text-to-World

Carpenters work with their hands. What other careers require people to work with their hands?

Glossary

apprenticeship – an arrangement in which someone learns a trade under an expert.

client – a person who hires or uses the services of a professional of some type.

construction industry – the area of work relating to building, repairing, renovating, and maintaining structures.

frame – a structure made of parts that are joined together and that supports a larger object.

install – to put into position and make ready for use.

molding – a strip of wood, stone, or other material used to frame or finish a door, window, or wall.

tradesperson – a person who works in an occupation that requires manual skill.

Index

Online Resources

popbooksonline.com

Thanks for reading this Cody Koala book!

This book is filled with videos, puzzles, games, and more! Scan the QR codes* while you read, or visit the website below to make this book pop.

popbooksonline.com/carpenter

*Scanning QR codes requires a web-enabled smart device with a QR code reader app and a camera.